SPORTS MATH

BY ERIN SULLIVAN

Table of Contents

Where's the Math in Sports?

You may be wondering what math has to do with sports. Think about shooting baskets, running in a race, or even doing tricks on a skateboard. Are you ever using math when you do these things?

The answer is yes! Have you ever kept track of the points you scored during a basketball game? Have you ever timed your running **speed?** Have you ever tried to do a "180" on your skateboard?

Even when you're just sitting in the stands at a sports event, you're using math—probably without even thinking about it. You need to use math to keep track of your team's wins and losses. You need to use math to see how much time is left in a game. You even need to use math to see if you have enough money to buy a hot dog! Whether you're an athlete or a fan, math comes in handy.

Team Sports

People love to play and watch team sports. One of the most popular team sports is baseball.

Some baseball fans are really into numbers. They can tell you how many times a player has struck out, how many games a team has won, and even how many home runs have been hit in a particular baseball park. More than anything else, though, baseball fans seem to enjoy arguing over who the greatest baseball player of all time was. A strong batter? A fast runner? A good fielder? An awesome pitcher? What do you think?

Nolan Ryan was a great pitcher.

Many people think that Babe Ruth was baseball's greatest hitter. His lifetime batting average was .342.

One way to pick a baseball great is to look at the player's **batting average**. A batting average tells how often a player gets a hit compared to how often he bats.

Hitting a ball in the major leagues is harder than you might think. Baseball great Ted Williams once said that the best players hit the ball only three out of every ten times at bat!

Ted Williams was a great batter.

Solve This

PROBLEM 1

Here's how you can find a player's batting average:
number of hits ÷ number of times at bat = batting average
Using a calculator, find the following players' batting averages. Round to the nearest thousandth.

PLAYER	HITS	AT BATS	BATTING AVERAGE
Jim	16	64	
Stacey	21	76	
José	14	55	
Lena	19	60	

CHAPTER 2

This photo from 1935 shows football players practicing.

Football is one of the great American sports. Some people think that football began back in the 1860s, when college students played a new version of **rugby.** Both football and rugby are games in which players have to get the ball into the goal by running with it, throwing it, and kicking it.

This stadium at the University of Wisconsin holds about 80,000 people!

Ever since the first football game was played, fans have been coming out to watch their teams compete. While high school fields might have room for only 100 fans, college and professional football teams have room for thousands of people.

Most teams keep track of **attendance** at each game. Attendance is the number of fans who come to watch the game. Does each team send out a person to count all of the people in their seats? Probably not. Instead, officials keep track of how many tickets are sold.

Solve This

PROBLEM 2

This chart shows the attendance at five of the Baltimore Ravens' home games.

Round each attendance number to the nearest hundred.

GAME	ATTENDANCE	ROUNDED TO THE NEAREST HUNDRED
1	68,843	
2	68,481	
3	69,200	
4	69,405	
5	69,416	

Michael Jordan scores a 2-pointer.

It is said that basketball was invented in Massachusetts on a cold winter day back in 1891. A teacher named James Naismith wanted to invent a team sport that his students could play inside. He got a soccer ball and hung two peach baskets at either end of the gym. During that very first game, Naismith needed a ladder to get the ball out of the basket every time a team scored!

Since then, basketball has become a favorite sport. And in 1997, the world of basketball took a big step forward when the Women's National Basketball Association (WNBA) was started.

Women's and men's basketball games follow most of the same rules. Each team has five players on the court. They dribble and pass the ball down the court to shoot baskets. Different kinds of baskets earn different points.

Depending on where the player is standing when making the shot, a shot can be worth one point, two points, or three points.

Sheryl Swoopes was the first woman signed to the WNBA in 1997.

Solve This

PROBLEM 3

Let's say you shoot two 3-pointers, four 2-pointers, and two 1-pointers in one game.
Your math problem would look like this:
(2 x 3) + (4 x 2) + (2 x 1) = 16 points

Use the information in this chart to find out how many points each player scored.

Remember to multiply inside the parentheses first and then add!

PLAYER	1 POINT	2 POINTS	3 POINTS	TOTAL
Leah	4	3	2	
Morgan	3	1	1	
Dawn	5	3	1	
Lisa	2	0	5	

CHAPTER 3

Individual Achievement

More than 2,500 years ago, the Greeks fought a long and difficult battle against their enemies, the Persians. When the Greeks finally won, they sent a runner from the site of their victory at Marathon to Athens, a distance of about twenty-six miles, to announce the good news.

Today, people compete in **marathons**, which are running races named in honor of this amazing trip. The first official marathon was run in 1896, at the very first modern **Olympics**. In addition to the Olympic marathon, cities all over the world hold marathons every year.

Like the original marathon, the modern-day marathon is about 26 miles long: 26 miles and 385 yards, to be exact! Record-setting runners often finish the race in a little over two hours. That's equal to running each mile in about five minutes!

Most marathons today have a category for wheelchair racers.

The Boston Marathon is held every year in April.

Solve This

PROBLEM 4

A marathon is 26 miles, 385 yards. Remember that there are 3 feet in a yard and 5,280 feet in a mile.

- How many feet are in a marathon?
- How many yards are in a marathon?
- How many inches are in a marathon?

Solve This

PROBLEM 5

Some Olympic pools are 50 meters long.
How many lengths of the pool would you need to swim in each of these events?

50 meters	
100 meters	
200 meters	
400 meters	
800 meters	

At the Summer Olympics, there are 32 different swimming events, 16 for men and 16 for women. Some races are 50 meters long, while others are 1,500 meters.

A swimmer in a short race will have to swim hard and fast from the beginning, while a swimmer in a longer race will try to save some energy for the end.

This swimmer is doing the butterfly stroke.

This diagram shows from top to bottom three steps of a flip turn.

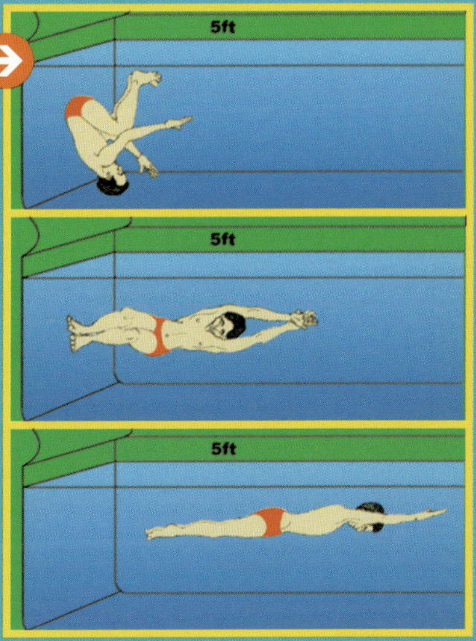

In every swimming race, the swimmer's goal is to finish in the fastest time. Many swimmers focus on making their starts and their turns as quickly as possible. Often, a swimmer who has a slow start or makes a slow turn has a hard time catching up with the leader.

Solve This

PROBLEM 6

Here's how you calculate a swimmer's speed: distance traveled ÷ time passed = speed. Let's say you swam 50 meters (m) in 25 seconds (sec). Your math problem would look like this: 50 m ÷ 25 sec = 2 m/sec. This tells you that you swam 2 meters per second!

Use a calculator to find out how fast these athletes swam.

Round each answer to the nearest hundredth. Remember to turn minutes into seconds!

SWIMMER	DISTANCE	TIME	SPEED (m/sec)
Diana	100 meters	1 min, 10 sec	
Lee	200 meters	2 min, 6 sec	
Joanna	400 meters	4 min, 18 sec	

balance beam

Gymnastics is one of the most popular Olympic sports. There are many different kinds of events for gymnasts to compete in, such as balance beams, parallel bars, and rings. For some fans, a favorite event is the floor exercise, where gymnasts do jumps, tumbles, and handsprings to music.

parallel bars

Gymnasts must be very strong and very flexible. They must also have an excellent sense of balance.

Most gymnasts begin serious training at a very early age, around four or five. After a lot of practice, some gymnasts enter local competitions. If they do well, they go on to state and national competitions. If they are lucky, they might get to try out for the Olympic team. Only fifteen out of thousands get to be part of the American team.

These young gymnasts are practicing balance.

Solve This

PROBLEM 7

This chart shows the number of medals won by each country in gymnastic events during an Olympics.

Use these data to create a bar graph.

COUNTRY	MEDALS WON
Russia	20
China	8
Romania	6
Latvia	1
Hungary	1
Spain	1
Ukraine	3
France	2
Belarus	2
Greece	2

Extreme Sports

What's an extreme sport? If you ask an extreme athlete, he or she might tell you that it's a sport that's fun, exciting, and a little risky.

Skateboarding is an extreme sport. It first became popular back in the 1950s after surfing had gained many fans. Back then, people compared skateboards to surfboards. Skateboarders were called "sidewalk surfers" because of the way they glided over the ground.

Skateboarding has changed a lot since its early days.

Solve This

PROBLEM 8

A circle is 360°. When skateboarders do a trick called a 360 they complete one full turn.

How many turns does a skateboarder complete for each trick listed below?

DEGREES	TURNS
180	
360	
540	
720	
900	
1080	

Many skateboarders do tricks called 360s and 180s. A 360 is a complete turn in the air. A 180 is a half-turn. These tricks get their names from the way people measure circles, in **degrees.**

Imagine the hands of a clock. If the minute hand moves halfway around the circle from the 12 to the 6, it has moved 180 degrees. If the minute hand moves all the way around the clock in a full circle, that's 360 degrees.

Think it over!

Some people think that the first skateboard was born when someone's scooter broke, and the rider just kept going without the handlebars!

Tony Hawk was the first person to do a 900. That's 360 + 360 + 180, or 2 ½ full turns in the air!

Mountain biking is another popular extreme sport. Instead of riding along flat, paved roads, mountain bikers head for hills and forests. Mountain bikers enjoy the excitement of riding on dirt trails.

There are two kinds of mountain biking. In downhill biking, bikers choose steep, downhill slopes. Downhill biking competitions are a lot like skiing events, and riders cruise down difficult courses at top speeds. Cross-country bikers race on flatter surfaces. They also ride at high speeds, but their trails include uphill slopes and other obstacles.

Mountain bikers choose downhill or cross-country events.

Mountain bikes are different from regular road bikes. One difference is in the bike's **frame**, or main body of the bike. A mountain bike's frame is heavier so that it can handle steep, difficult trails.

Mountain bikes also have thicker tires than a regular bike. The thicker tires make the bikes more stable. If a rider hits a rock or a bumpy patch on the trail, the bike is less likely to tip over.

Solve This

PROBLEM 9

It's your lucky day! You've won $1,000 in a bike race.

What is the total cost of buying new mountain biking equipment? How much will you have left over?

Bike	$ 469.00
Windbreaker pants and jacket	$ 120.00
Helmet	$ 27.00
Bicycle Pump	$ 25.00
Bike Lock	$ 35.00
Bike Mirror	$ 17.00
Repair Kit	$ 30.00

If you like problem solving, hard climbing, and high places, then rock climbing is the sport for you. Rock climbers are always looking for the next challenge. It might be a rocky wall in the park or a steep cliff in the wild. It might even be the wall of a rock-climbing gym.

Most rock climbers use special tools to keep them safe. Climbers use special ropes to rope themselves together. If a climber slips and misses a foothold, the ropes will keep him or her from falling too far.

Climbers have special shoes that help them get toeholds in the rock. They also use bolts that they drive into the rock so that they can hold on more easily.

Serious rock climbers use special equipment to stay safe.

When you're thousands of feet up in the air, gripping the rock with your hands and feet, you might start feeling a little nervous. But expert climbers say that this is the most important time to relax! Climbers need to stay calm while they think about where to put their hands and feet next.

These climbers are practicing at an indoor gym.

Solve This

PROBLEM 10

This is a list of ten of the tallest mountains in the world.

Put them in order from tallest to shortest.

PEAK	
Kangchenjunga	28,208
Makalu I	27,790
Cho Oyu	26,750
K2	28,250
Manaslu I	26,760
Nanga Parbat	26,660
Everest	29,035
Annapurna I	26,504
Lhotse	27,890
Dhaulagin I	26,810

Unusual Sports

All over the world, people take part in many kinds of unusual sports. You may not know a lot about these sports.

One of the oldest and most popular sports in Japan is **sumo wrestling.** Sumo wrestlers compete in a ring that is eighteen feet square, with a circle inside that is fifteen feet across. Above the ring is a special roof, with four different-colored tassels hanging from the corners. These represent the seasons of the year.

The sumo wrestling ring is called a *dohyo* (DOH-yoh). It is made of clay and lightly covered with sand. Japanese sumo wrestlers are called *rikishi* (REE-kih-shee).

When two wrestlers first meet in the ring, they stare at each other. They are trying to make each other nervous. Then they jump into action! A wrestler loses when any part of his body (other than his feet) touches the ground. The actual wrestling part of each match may last only a few seconds!

The best wrestlers in Japan are highly respected for their strength, their bravery, and their size.

Solve This

PROBLEM 11

Use this chart to find the average weight of these sumo wrestlers.

Sumo Wrestler	Weight
Musashimaru	483 lbs
Takanohana	351 lbs
Musoyama	392 lbs
Kaio	384 lbs
Dejima	359 lbs
Miyabiyam	395 lbs

Back in 1925, a terrible thing happened in Nome, Alaska. A disease called **diphtheria** (dif-THEER-ee-ah) spread through the tiny town. Many children got sick. And the medicine they needed was far away, in Anchorage.

The weather was so bad that no planes could fly into Nome. Instead, people in Anchorage decided to use dogsleds and drivers called **mushers** to get the medicine there. Twenty different dogsled teams and mushers worked together to deliver the medicine to Nome in only five and a half days.

Every year, there is a dogsled race from Anchorage to Nome to commemorate the bravery of the very first mushers. It is called the **Iditarod** (eye-DIH-ter-ahd).

The Iditarod trail runs for about 1,150 miles. Mushers drive teams of dogs across the frozen ground and stop at the tiny villages along the way to eat and rest. It is an extremely difficult race, taking place in a cold, windy, and isolated area.

Balto was the dog who led the final dogsled team into Nome. This is his statue in New York City.

Winning teams usually take nine or ten days to finish the Iditarod.

Solve This

PROBLEM 12

1. One dogsled team traveled 60 miles on Monday, 75 miles on Tuesday, and 49 miles on Wednesday. How many miles did it travel in all?

2. If this same team traveled 260 miles by the end of Friday, how far did it travel on Thursday and Friday?

3. How many more miles does it have to travel to finish the race?

How many times can you say "kabaddi" without taking a breath? Have a friend count your kabaddis and record the data on a class chart. Work with your classmates to create a bar graph that shows the class's kabaddi count!

In India, people play an unusual version of tag called **kabaddi** (ku-BUH-dee). In this game, there are two teams with seven players each. One team sends a player over to the other side to tag the opposite team's players. While the tagger is running, he or she must say the word "kabaddi" over and over again without taking a breath.

If the runner gets back to his or her own side without taking a breath, then the tagged players are out, and the runner's team gets points. If the runner takes a breath, he or she is out!

Kabaddi is one of India's most popular sports. Even though it sounds like a kids' game, adults play it, too. In fact, there are professional kabaddi teams that compete across the country. There is even a world championship, in which teams from Pakistan, China, Canada, and the United States compete.

Kabaddi is over 4,000 years old!

Sports Math at Work

By now you probably understand that math and sports really are related. You use math to keep track of your progress while you train for a sport. You use math when you buy the equipment that you need. You use math when you keep track of the time and count up the points during games.

Using math during sports can help you become a better athlete and a better fan. When you start to think about it, you'll probably come up with other ways of using math during sports that we haven't even mentioned in this book.

But the best part about sports math is that it's a lot of fun. The next time you go outside to play, think about all the math work your brain is doing!

Solve This Answers

1 Page 5
Jim: .250
Stacey: .276
José: .255
Lena: .317

2 Page 7
1: 68,800
2: 68,500
3: 69,200
4: 69,400
5: 69,400

3 Page 9
Leah: 16
Morgan: 8
Dawn: 14
Lisa: 17

4 Page 11
In one marathon, there are 138,435 feet; 46,145 yards; 1,661,220 inches

5 Page 12

50 meters	1
100 meters	2
200 meters	4
400 meters	8
800 meters	16

6 Page 13
Diana: 1.43 m/sec;
Lee: 1.59 m/sec;
Joanna: 1.55 m/sec

7 Page 15
Bar Graph

8 Page 16
180: 1/2; 360: 1; 540: 1 1/2;
720: 2; 900: 2 1/2; 1080: 3.

9 Page 19
Total cost: $723.00
Change: $277.00

10 Page 21
Everest, K2, Kangchenjunga, Lhotse, Makalu I, Dhaulagin I, Manaslu I, Cho Oyu, Nanga Parbat, Annapurna I.

11 Page 23
Average weight: 394 pounds

12 Page 25
1. 60 + 75 + 49 = 184 miles
2. 260 - 184 = 76 miles
3. 1,150 - 260 = 890 miles

Glossary

attendance	the number of people who go to a game
batting average	the number that tells how many hits a batter gets, compared to the number of times at bat
degrees	units of measure for circles
diphtheria	a contagious disease that can cause damage to the heart and nervous system
frame	the body of a bike
Iditarod	a dogsled race in Alaska
kabaddi	a game of tag played mostly in India
marathon	a running race about twenty-six miles long
musher	a dogsled driver
Olympics	worldwide sports competitions
rugby	a game like football in which players try to get a ball into the goal using their hands and feet
speed	how fast an athlete is going: distance ÷ time
sumo wrestling	a kind of wrestling popular in Japan

Index